Julius A Dresser, Horatio W. (Horatio Willis) Dresser

The true history of mental science

The facts concerning the discovery of mental healing

Julius A Dresser, Horatio W. (Horatio Willis) Dresser

The true history of mental science
The facts concerning the discovery of mental healing

ISBN/EAN: 9783742829993

Manufactured in Europe, USA, Canada, Australia, Japa

Cover: Foto ©Thomas Meinert / pixelio.de

Manufactured and distributed by brebook publishing software
(www.brebook.com)

Julius A Dresser, Horatio W. (Horatio Willis) Dresser

The true history of mental science

THE TRUE HISTORY OF MENTAL SCIENCE

THE FACTS CONCERNING THE DISCOVERY OF MENTAL HEALING

BY

JULIUS A. DRESSER

REVISED, WITH NOTES AND ADDITIONS, BY
HORATIO W. DRESSER

AUTHOR OF "METHODS AND PROBLEMS OF SPIRITUAL HEALING," ETC.

Price 20 cents

BOSTON
GEO. H. ELLIS, 272 CONGRESS STREET
1899

PREFACE TO FIRST EDITION.

THE facts given in the following address * have been held for many years, until there should be such a general demand for them that they would receive a willing and unprejudiced ear, and an appreciation in accordance with their merits. Portions of this history have been contributed by the author to the *Mental Healing Monthly* of Boston, in nearly the same words as here given, and also some portions to the *Christian Metaphysician* of Chicago. But this pamphlet includes extracts from the unpublished manuscripts of P. P. Quimby, which appear in no other publication.

J. A. D.

BOSTON, May, 1887.

* Delivered at the Church of the Divine Unity, Boston, Feb. 6, 1887.

THE FACTS CONCERNING THE DISCOVERY OF MENTAL HEALING.

THE foundation principles of what we now term Mental Science are shown by history to have been largely understood by the philosophers of all ages. The philosophy of Plato, who flourished four hundred years before Christ, was essentially idealistic ; and the same idealism is set forth in different forms by leading thinkers of succeeding generations, notably by Bishop Berkeley, in the seventeenth century. But, while these thinkers mainly agreed that all reality is in the realm of spirit, of which what we see is only an emanation or manifestation, they all failed to apply their philosophy to the healing of disease. Of those who had this understanding previous to our time, only Jesus and his disciples applied it to relieving human ills. All others devoted their teachings simply to modifying and forming character, and to the development of a system of metaphysics.

Coming down to the nineteenth century, we find Ralph Waldo Emerson's writings permeated with the idealistic theory, and running over with the belief in an omnipresent Goodness as the substance of all things, with here and there a hint that the so-called "ills that flesh is heir to" may be eradicated as errors, when held up to the light of truth, the same as can moral evils. But the first person in this age who penetrated the depths of truth so far as to discover and bring forth a true science of life, and publicly apply it to the healing of the sick, was Phineas Parkhurst Quimby, of Belfast, Me.* I am well aware that with some people this is a disputed point, as they respect the claims of certain others; but I have been requested, by many persons interested in this history, to say what I know about it. Believing that the time has come to do so, I shall now give you a series of facts, and you can judge of them for yourselves.

The first that I knew of P. P. Quimby was in June, 1860, when I went to him as a patient, in Portland, Me. This was five and a half years before his death. He had then, 1860, been in the regular practice of mental healing for many years, in different towns in Maine, and had been located in Portland about two years. There was at that time, 1860, no one else in the practice in New England, nor in this country, so far as was then known, or has since been heard of; nor was there at that time any one else who understood it as a science, he having been the discoverer and founder, as I think I shall show you.† He had then, 1860,

---

* Born in Lebanon, N.H., February 16, 1802. Died in Belfast, Me., January 16, 1866.

† Perhaps the nearest approach to the theory and practice of Dr. Quimby was the contemporary investigation of John Bovee Dods, who believed that

been at work more than twenty years in this field of discovery and practice, which carries his first investigations previous to the year 1840.

The question may be asked, "Was Quimby ever a mesmerist?" I reply that he was, for a limited time, and for purposes of experiment and investigation. The truth came to him, not as a revelation pure and simple, but as the result of practical experiment and patient research among the phenomena of life, urged on by the impulses of an active, inquiring, comprehensive mind. I have seen extracts from newspapers as far back as 1842–43, giving accounts of his public exhibitions of mesmerism, in some of which newspaper accounts he was rated with a few others in this country and Europe who were the leading mesmerizers in the world. Dr. Quimby had been a watch and clock maker for some years, when mesmerism attracted his attention.

The subject of mesmerism was first introduced into this country by Mr. Charles Poyan, a French gentleman, in the year 1836. A few years later a certain Dr. Collyer lectured upon it in New England and elsewhere. In 1840 P. P. Quimby commenced experimenting with it, although this did not furnish him his first lesson in the truth he afterward developed, as I have learned from

electricity was the connecting link between mind and matter, that disease originates in the electricity of the nerves, and can be cured by a change of mind. See "The Philosophy of Electrical Psychology." New York, S. R. Wells, 1870.

accounts of his earlier experiences. From a newspaper account of one of his public exhibitions of mesmerism in Belfast, Me., dated April 27, 1843, I make this excerpt:—

"Before we proceed to describe the experiments" (the newspaper says), "we will say that Mr. Quimby is a gentleman, in size rather smaller than the medium of man, with a well-proportioned and well-balanced phrenological head, and with the power of concentration surpassing anything we have ever witnessed. His eyes are black and very piercing, with rather a pleasant expression; and he possesses the power of looking at one object, without even winking, for a great length of time."

In his mesmeric experiments, as reported in the Maine papers in those years so long ago, Quimby is shown to have progressed gradually *out* of mesmerism into a knowledge of the hidden powers of mind; and he soon found in man a principle, or a power, that was not of man himself, but was higher than man, and of which he could become a medium. Its character was goodness and intelligence, and its power was great. He also found that disease was primarily an erroneous belief of mind. Here was a discovery of truth; and on this discovery he founded a system of treating the sick, and founded a "science of life." As a better testimony than my own on these points, I will here introduce a quotation from a letter by his son, George A. Quimby, of Belfast, Me., in reply to one in which I wrote for certain data relating to his father. Speaking of his father, Mr. Quimby wrote as follows : —

"Some time in 1840 he became deeply interested in mesmerism ; and for quite a number of years, in connection with his other business, he gave exhibitions with a clairvoyant subject through the state of Maine, and also treated disease, using his mesmeric power, as it was termed then. This method he kept up, but gradually working *out* of the mesmeric idea into a train of reasoning of his own, which he applied to the patient, till finally he gave up putting the patient to sleep mesmerically,* and followed the mode of treatment which he originated and continued up to the time of his death, which treatment, in his own words, was this : —

"He says : 'My practice is unlike all medical practice. I give no medicine, and make no outward applications. I tell the patient his troubles, and what he thinks is his disease ; and my explanation is the cure. If I succeed in correcting his errors, I change the fluids of the system and establish *the truth, or health. The truth is the cure.* This mode of practice applies to all cases.'"

These are Dr. Quimby's own words, and any one can see that they mean a purely mental treatment ; for he speaks of what the patient thinks is

---

* For a more detailed account of this progress out of mesmerism see "The Philosophy of P. P. Quimby," by A. G. Dresser, Geo. H. Ellis, 1895, p. 16.

his disease, and calls it his error, by saying that, if he succeeds in correcting the patient's errors, he then establishes the truth, and *the truth is the cure.* You see from this that he had discovered that disease was an error of mind, and nothing else, and the God-power of truth which he had discovered in man, being set up again in the victim of disease, destroyed the error, or disease, and re-established the harmony.

This discovery, you observe, was not made from the Bible, but from mental phenomena and searching investigations; and, after the truth was discovered, he found his new views portrayed and illustrated in Christ's teachings and works. If you think this seems to show that Quimby was a remarkable man, let me tell you that he was one of the most unassuming of men that ever lived; for no one could well be more so, or make less account of his own achievements. Humility was a marked feature of his character (I knew him intimately). To this was united a benevolent and an unselfish nature, and a love of truth, with a remarkably keen perception. But the distinguishing feature of his mind was that he could not entertain an opinion, because it was not knowledge. His faculties were so practical and perceptive that the wisdom of mankind, which is largely made up of opinions, was of little value to him. Hence the charge that he was not an educated man is literally true. True knowledge to him was *positive proof,* as in a problem of mathematics. Therefore, he discarded books and sought phenomena, where his perceptive faculties made him master of the situation. Therefore, he got from his experiments in mesmerism what other men did not get,—a stepping-stone to a higher knowledge than man possessed, and a new range to mental vision. He wrote out his discoveries at great length; and from these yet unpublished writings, now in the possession of his son, before referred to, I am privileged to incorporate in this lecture the following article, which was written in the year 1863, and thus allow Quimby to tell an important part of his own story. These are his words:—

"MY CONVERSION FROM DISEASE TO HEALTH, AND THE SUBSEQUENT CHANGE FROM BELIEF IN THE MEDICAL FACULTY TO ENTIRE DISBELIEF IN IT, AND TO THE KNOWLEDGE OF THE TRUTH ON WHICH I BASE MY THEORY."

"Can a theory be found, capable of practice, which can separate truth from error? I undertake to say there is a method of reasoning which, being understood, can separate one from the other. Men never dispute about a fact that can be demonstrated by scientific reasoning. Controversies arise from some idea that has been turned into a false direction, leading to a false position. The basis of my reasoning is this point: that whatever is true to a person, if he cannot prove it, is not necessarily true to another. Therefore, because a person says a thing is no reason that he says true. The greatest evil that follows taking an opinion for a truth is disease. Let medical and religious opinions, which produce so vast an amount of misery, be tested by the rule I have laid down, and it will be seen how much they are founded in truth. For twenty years I have been testing them, and I have failed to find one single principle of truth in either. This is not from any prejudice against the medical faculty; for, when I began to investigate the mind, I was entirely on that side. I was prejudiced in favor of the medical faculty; for I never employed any one outside of the regular faculty, nor took the least particle of quack medicine.

" Some thirty years ago I was very sick, and was considered fast wasting away with consumption. At that time I became so low that it was with difficulty I could walk about. I was all the while under the allopathic practice, and I had taken so much calomel that my system was said to be poisoned with it; and I lost many of my teeth from that effect. My symptoms were those of any consumptive; and I had been told that my liver was affected and my kidneys were diseased, and that my lungs were nearly consumed. I believed all this, from the fact that I had all the symptoms, and could not resist the opinions of the physician while having the proof with me. In this state I was compelled to abandon my business; and, losing all hope, I gave up to die, — not that I thought the medical faculty had no wisdom, but that my case was one that could not be cured. "Having an acquaintance who cured himself by riding horseback, I thought I would try riding in a carriage, as I was too weak to ride horseback. My horse was contrary; and once, when about two miles from home, he stopped at the foot of a long hill, and would not start except as I went by his side. So I was obliged to run nearly the whole distance. Having reached the top of the hill I got into the carriage; and, as I was very much exhausted, I concluded to sit there the balance of the day, if the horse did not start. Like all sickly and nervous people, I could not remain easy in that place; and, seeing a man ploughing, I waited till he had ploughed around a three-acre lot, and got within sound of my voice, when I asked him to start my horse. He did so, and at the time I was so weak I could scarcely lift my whip. But excitement took possession of my senses, and I drove the horse as fast as he could go, up hill and down, till I reached home; and, when I got into the stable, I felt as strong as I ever did." (The account of this experience ends here; but it seems to have been a phenomenon that opened Quimby's eyes, and through his keen perceptions it taught him a great deal. The article continues with an experience of another nature.

"When I commenced to mesmerize, I was not well, according to the medical science; but in my researches I found a remedy for my disease. Here was where I first discovered that mind was matter, and capable of being changed.

"Also that, disease being a deranged state of mind, the cause I found to exist in our belief. The evidence of this theory I found in myself; for, like all others, I had believed in medicine. Disease and its power over life, and its curability, are all embraced in our belief. Some believe in various remedies, and others believe that the spirits of the dead prescribe. I have no confidence in the virtue of either. I know that cures have been made in these ways. I do not deny them. But the principle on which they are done is the question to solve; for disease can be cured, with or without medicine, on *but one principle.* I have said I believed in the old practice and its medicines, the effect of which I had within myself; for, knowing no other way to account for the phenomena, I took it for granted that they were the result of medicine.

"With this mass of evidence staring me in the face, how could I doubt the old practice? Yet, in spite of all my prejudices, I had to yield to a stronger evidence than man's opinion, and discard the whole theory of medicine, practiced by a class of men, some honest, some ignorant, some selfish, and all thinking that the world must be ruled by their opinions.

"Now for my particular experience. I had pains in the back, which, they said, were caused by my kidneys, which were partially consumed. I also was told that I had ulcers on my lungs. Under this belief, I was miserable enough to be of no account in the world. This was the state I was in when I commenced to mesmerize. On one occasion, when I had my subject asleep, he described the pains I felt in my back (I had never dared to ask him to examine me, for I felt sure that my kidneys were nearly gone); and he placed his hand on the spot where I felt the pain. He then told me that my kidneys were in a very bad state, —that one was half-consumed, and a piece three inches long had separated from it, and was only connected by a slender thread. This was what I believed to be true, for it agreed with what the doctors told me, and with what I had suffered; for I had not been free from pain for years. My common sense told me that no medicine would ever cure this trouble, and therefore I must suffer till death relieved me. But I asked him if there was any remedy. He replied, 'Yes, I can put the piece on so it will grow, and you will get well.' At this I was completely astonished, and knew not what to think. He immediately placed his hands upon me, and said he united the pieces so they would grow. The next day he said they had grown together, and from that day I never have experienced the least pain from them.

"Now what is the secret of the cure? I had not the least doubt but that I was as he described; and, if he had said, as I expected that he would, that nothing could be done, I should have died in a year or so. But, when he said he could cure me in the way he proposed, I began to think; and I discovered that I had been deceived into a belief that made me sick. The absurdity of his remedies made me doubt the fact that my kidneys were diseased, for he said in two days they were as well as ever. If he saw the first condition, he also saw the last; for in both cases he said he could see. I concluded in the first instance that he read my thoughts, and when he said he could cure me he drew on his own mind; and his ideas were so absurd that the disease vanished by the absurdity of the cure. This was the first stumbling-block I found in the medical science. I soon ventured to let him examine me further, and in every case he would describe my feelings, but would vary about the amount of disease; and his explanation and remedies always convinced me that I had no such disease, and that my troubles were of my own make.

"At this time I frequently visited the sick with Lucius, by invitation of the attending physician; and the boy examined the patient, and told facts that would astonish everybody, and yet every one of them was believed. For instance, he told a person affected as I had been, only worse, that his lungs looked like a honeycomb, and his liver was covered with ulcers. He then prescribed some simple herb tea, and the patient recovered ; and the doctor believed the medicine cured him. But I believed that the doctor made the disease ; and his faith in the boy made a change in the mind, and the cure followed.* Instead of gaining confidence in the doctors, I was forced to the conclusion that their science is false. Man is made up of truth and belief ; and, if he is deceived into a belief that he has, or is liable to have, a disease, the belief is catching, and the effect follows it. I have given the experience of my emancipation from this belief and from confidence in the doctors, so that it may open the eyes of those who stand where I was. I have risen from this belief ; and I return to warn my brethren, lest, when they are disturbed, they shall get into this place of torment prepared by the medical faculty. Having suffered myself, I cannot take advantage of my fellow-men by introducing a new mode of curing disease and prescribing medicine. My theory exposes the hypocrisy of those who undertake to cure in that way. They make ten diseases to one cure, thus bringing a surplus of misery into the world, and shutting out a healthy state of society. They have a monopoly, and no theory that lessens disease can compete with them. When I cure, there is one disease the less ; but not so when others cure, for the supply

* See a more adequate statement of this theory, " The Philosophy of P. P. Quimby," p. 90.

of sickness shows that there is more disease on hand than there ever was. Therefore, the labor for health is slow, and the manufactory of disease is greater. The newspapers teem with advertisements of remedies, showing that the supply of disease increases. My theory teaches man to manufacture health ; and, when people go into this occupation, disease will diminish, and those who furnish disease and death will be few and scarce."

This account from Dr. Quimby himself settles many things. First, it gives in detail *one* of the *many* experiences by which he discovered this truth. It shows, also, the practical nature of the man's mind, and illustrates his wonderful perceptive powers. And the article shows that no one could have written it but the one whose experience it describes ; and it shows, too, that what he arrived at was the knowledge that disease is nothing but an error of belief, to be corrected by the truth. On this basis he practiced ever afterward. How could he do otherwise, after making such a discovery? And this discovery was made about forty-five years ago. All these facts can be fully substantiated by consulting certain newspaper files,* and certain persons who are familiar with it all. And this theory, that disease is an error of belief to be corrected by the truth, not only formed the basis of "a science of health" which Dr. Quimby introduced, but it is the sub-

* Some of these newspaper articles have been reprinted in "The Philosophy of P. P. Quimby," p. 22.

ject of voluminous manuscripts devoted to the "true science of life and happiness," and others in which he explained and defended Christ's sayings, his gospel and his work. He also wrote upon the true standard of law, and of government, and upon other topics. All these writings I have read, being in the confidence of George A. Quimby, the son, who holds them. This son was with his father as secretary during the father's last five years of practice, and the father hoped that his son would take up the practice with him, and succeed him ; but he took a different turn, and is a manufacturer.

I think I can see a wisdom in nearly everything. If these writings had been published, as Dr. Quimby intended, in his day, or even at any time since, they would have found a public unprepared for them. Therefore, they are in the hands of a person whose sympathies are not stirred by a work in the truth, as some of ours are, to issue them before their time. But these manuscripts will be published at a future day. The present owner of them is not troubled in the least, nor am I, by such misstatements, to call them by no worse name, as have appeared in certain recent publications, belying and belittling Dr. Quimby and your speaker; nor by the efforts to show that Dr. Quimby's manuscripts were written by somebody else.* We have only pity for a person who would

* It has frequently been claimed that Mrs. Eddy was Dr. Quimby's secretary, and that she helped him to formulate his ideas. It has also been stated

make such misstatements for a purely selfish purpose. You have heard or read of Haman's gallows, which he built for Mordecai, the Jew, and how they finally hung only the builder himself! P. P. Quimby's writings, when published, will speak for themselves; and his friends know it perfectly. They pay no attention to what is said or done by others, any more than your speaker does.

One more point of a personal nature, and I pass on to more general facts. I have been blamed many times by persons for so long withholding the facts I now make public, together with much more that I shall not mention at this time. The reply that myself and wife have always made was that we wished to pursue our work in the truth in the true spirit of that truth, which is charity and love; and this did not lead us to say aught against another, nor even to reply to unjust aspersions upon ourselves or P. P. Quimby. We knew that some time the facts would be called for by sup-

that these manuscripts were Mrs. Eddy's writings, left by her in Portland; that the articles printed in this pamphlet were Mrs. Eddy's words, as nearly as she can recollect them (*Christian Science Sentinel*, Feb. 16, 1899). *There is absolutely no truth* in any of these statements or suppositions. Mrs. Eddy never saw a page of the *original* manuscripts; and volume 1, loaned her by my father in 1862, was his *copy from a copy*. Mrs. Eddy may have made a copy of this volume for her own use, but the majority even of the copied articles Mrs. Eddy never saw. I have read and copied all of these articles, and can certify that they contain a very original and complete statement of the data and theory of mental healing. There are *over eight hundred closely written pages*, covering one hundred and twenty subjects, written previous to March, 1862, more than six months before Mrs. Eddy went to Dr. Quimby. The later articles are chiefly elaborations of these earlier essays. The ideas and methods of treatment were original with Dr. Quimby: the language of "Science and Health" is original with Mrs. Eddy.

porters of the truth; and then, if we felt the time had come, we would give them, and, in so doing, the statement of the facts would be seen plainly to be only for the truth's sake, and in its true spirit.

But, said some, people are being deceived constantly. We replied that we knew it, but that very many things were allowed by an overruling Providence to go on in this world which seemed to need correction, and we were not running this world nor the universe, and we were more particular to see that we discharged our own immediate duties and what little good we might do, to the best of our ability and in the spirit of love, than we were to follow the prevailing human custom of taking upon ourselves the assumed work of correcting errors that other people were falling into. I acknowledge that our course has been open to criticism on this point; but, dear friends, we have waited, "with malice toward none and charity for all," for your call for these facts; and now we can give them because they are wanted, and not because of any pleasure it might be said we had in doing it. For that phase of this history which constitutes unavoidable reflection upon others is only painful; and I have avoided it as much as possible in giving this account, and have therefore left out a *great deal* of detail that might otherwise have been included, and have made no replies at all to the many unjust and *false* statements that have been published and circulated about P. P. Quimby and myself. No one has been really injured as a result of our reticence, I trust; and certainly we have not suffered in any respect, except one. And that is that persons have said, and others have thought, that, because we made no replies to anything, there were none that we could make, and therefore all claims and statements of others were true. Do you remember the words of the apostle,— "Charity suffereth long, and is kind; is not easily provoked, thinketh no evil, rejoiceth not in iniquity, but rejoiceth in the truth; charity never faileth"?

Such is the spirit of the kind of truth that I learned from P. P. Quimby, and the kind that he himself practiced; and his spirit of love so opened his soul to the God-power that his works were marvellous. The quick cures that he brought about have not been equalled by any one since his time, so far as I know. Myself and wife have owed our lives to him for nearly twenty-seven years past, and to the truth he revealed to us. Thousands of others could make a similar testimony, but I prefer not to occupy time with relating his cures. The man himself never desired publicity. The truth itself and the good of humanity were the first and last considerations with him. He even had no fixed name for his theory * or practice, desiring to be known only by his fruits. He sank

* Dr. Quimby frequently termed his theory "The Science of Life and Happiness," and in one article he uses the term "Christian Science."

the individual wholly in the cause of truth and the good of humanity.

It is the intention of your speaker to relate this history so as to avoid any appearance of fulsome praise, because the man Quimby would not desire it; and it is my aim only to relate plain facts in a plain manner, and I request you therefore to consider no statement herein as overdrawn. Your attention is called to one important fact; and that is, that the kind of individual I am describing in the person of P. P. Quimby is the kind who *can make discoveries of truth,* if any one can,—that is, a mind of great capabilities, coupled with great humility and extreme unselfishness. This is the kind of a medium that God speaks through, because such a soul is open to his inspirations. On the other hand, a selfish soul, who seeks personal aggrandizement, is not open to revelations of much moment, because selfishness always blinds one. The truth does not flourish in such soil.

P. P. Quimby's perceptive powers were something remarkable. He always told the patient, at the first sitting, what the latter thought was his disease; and, as he was able to do this, he never allowed the patient to tell him anything about his case. Quimby would also continue and tell the patient what the circumstances were which first caused the trouble,* and then explain to him how he fell into his error, and then from this basis

---

* See "The Philosophy of P. P. Quimby," p. 51.

prove to him, in many instances, that his state of suffering was purely an error of mind, and not what he thought it was. Thus his system of treating diseases was really and truly a science, which proved itself. You see also from these statements, how he taught his patients to understand, and how persons who went to him for treatment were instructed in the truth as well as restored to health. In this way some persons became especially instructed, as did your speaker. The persons referred to also obtained many private interviews with him for further instruction in the truth.

The question has often been asked, by persons who had gotten some idea of the truth, how Dr. Quimby, who was a man of such power and understanding, came to die? Herein lies a story of his unselfishness. The man was overrun with patients for many years, and he was alone in the work. There is always a limit to finite endurance, and his heart was too large to enable him to refuse people whom he might help out of their sufferings when they applied to him. During those years when his office was in Portland, his home and family being always in Belfast, he was compelled once in four or six weeks to get away from the pressing tide of humanity, and go home to Belfast, privately, and rest for three or four days.

He sometimes would say to those nearest him that, if he ever should allow himself to get so far exhausted as not to be able to recover himself, there was no one to help him, and he would be compelled to pass out. But, though he never expected to overdo to that extent, it is just what happened. In brief, he laid down his life for the sick, and died in their cause, at the age of sixty-four years.

Nearly all, in those days, who were willing to try a practitioner outside of the medical schools, were persons who had exhausted every means of help within those schools; and, when finally booked for the grave, they would send for or go to Quimby. As he expressed it, they would send for him and the undertaker at the same time, and the one who got there first would get the case. Consequently, his battle with error, alone and single-handed, was a hard one, especially as in those days there was much less liberality than now.

Some may desire to ask if, in his practice, he ever in any way used manipulation. I reply that, in treating a patient, after he had finished his explanations, and the silent work, *which completed the treatment*, he usually rubbed the head two or three minutes, in a brisk manner, for the purpose of letting the patient see that something was done. This was a measure of securing the confidence of the patient at a time when he was starting a new practice, and stood alone in it.* I knew him to

* The following is an illustration of Mrs. Eddy's attempt to give these facts another interpretation (*Christian Science Journal*, June, 1887): "If, as Mr. Dresser says, Mr. Quimby's theory (if he had one) and practice were like mine,

make many and quick cures at a distance sometimes with persons he never saw at all. He never
considered the touch of the hand as at all necessary; but let it be governed by circumstances, as
was done eighteen hundred years ago.

But, dear friends, I do not wish you to rely
upon my statements alone for the facts of this
history, and of this man's character, his discoveries, and his works. To what has already been
given from the words of others, still further testi-
monies may be added. That able writer upon
Mental Science, Dr. W. F. Evans, pays the following tribute to Quimby in his second volume,
entitled "Mental Medicine." He says: "Disease
being in its root a *wrong belief*, change that
belief, and we cure the disease. . . . The late Dr.
Quimby, of Portland, one of the most successful
healers of this or any age, embraced this view of
the nature of disease, and by a long succession of
most remarkable cures . . . proved the truth of the
theory. . . . Had he lived in a remote age or country, the wonderful facts which occurred in his
practice would have now been deemed either

purely mental, what need had he of such physical means as wetting his hands in
water and rubbing the head? Yet these appliances he continued until he ceased
practice; and, in his last sickness, the poor man employed a homœopathic physician. The Science of Mind-healing would be lost by such means, and it is a
moral impossibility to understand or to demonstrate this science through such
extraneous aids. Mr. Quimby never, to my knowledge, taught that matter was
mind; and he never intimated to me that he healed mentally, or by the aid of
mind. Did he believe matter and mind to be one, and then rub matter in order
to convince the mind of truth? Which did he manipulate with his hands, matter
or mind? Was Mr. Quimby's entire method of treating the sick intended to
hoodwink his patients?"

mythical or miraculous. He seemed to reproduce the wonders of the gospel history." Dr. Evans obtained this knowledge of Quimby mainly when he visited him as a patient, making two visits for that purpose about the year 1863, an interesting account of which I received from him at East Salisbury in the year 1876. Dr. Evans had been a clergyman up to the year 1863, and was then located in Claremont, N.H. But so readily did he understand the explanations of Quimby, which his Swedenborgian faith enabled him to grasp the more quickly, that he told Quimby at the second interview that he thought he could himself cure the sick in this way. Quimby replied that he thought he could. His first attempts on returning home were so successful that the preacher became a practitioner from that time, and the result has been great growth in the truth and the accomplishment of a great and a good work during the nearly twenty-five years since then. Dr. Evans's six volumes * upon the subject of Mental Healing have had a wide and well-deserved sale.

Among those who were friends as well as patients of Quimby during the years from 1860 to 1865, † and who paid high tributes to his discoveries of truth, and the consequent good to many

---

* Published by H. H. Carter, Boston. The first volume appeared in 1869, six years before Mrs. Eddy's "Science and Health."

† Mrs. Eddy, then Mrs. Patterson, went to Portland as a patient in October, 1862, from Hill, N.H. She had been a confirmed invalid for six years, and, having heard of Dr. Quimby's "wonderful power," believed he could cure her.

people and to the world, was one who, for some strange reason, afterward changed and followed a different course, with which you all are more or less familiar. I refer to the author of "Science and Health." * As she had, during several years, special opportunities to know the man and to learn truth of him, this record would be incomplete without including her testimony at that time. Fortunately, it can be given in her own words; and you can form your own estimate of them.

When the lady became a patient of Quimby, she at once took an interest in his theory, and imbibed his explanations of truth rapidly. She also took a bold stand, and published an account of her progress in health in a daily paper. The following is an extract from her first article thus published, which appeared in the Portland *Evening Courier* in 1862 : —

"When our Shakespeare decided that 'there were more things in this world than were dreamed of in your philosophy,' I cannot say of a verity that

* It is claimed that Mrs. Eddy wrote this book by divine inspiration in 1866. It is therefore accepted by her followers as the supreme authority on mental healing, who compare Mrs. Eddy to Christ, and accept her misstatements concerning Dr. Quimby as absolute truth. In the *Christian Science Journal*, June, 1887, Mrs. Eddy says: "It was after the death of Mr. Quimby, and when I was apparently at the door of death, that I made this discovery in 1866. After that it took about ten years of hard work for me to reach the standard of my first edition of 'Science and Health,' published in 1875. As long ago as 1844 I was convinced that mortal mind produced all disease, and that the various medical systems were in no proper sense scientific. In 1862, when I first visited Mr. Quimby, I was proclaiming — to druggists, Spiritualists, and mesmerists — that science must govern all healing."

he had a foreknowledge of P. P. Quimby. And when the school Platonic anatómized the soul and divided it into halves, to be reunited by elementary attractions, and heathen philosophers averred that old Chaos in sullen silence brooded o'er the earth until her inimitable form was hatched from the egg of night, I would not at present decide whether the fallacy was found in their premises or conclusions, never having dated my existence before the flood. When the startled alchemist discovered, as he supposed, an universal solvent, or the philosopher's stone, and the more daring Archimedes invented a lever wherewithal to pry up the universe, I cannot say that in either the principle obtained in nature or in art, or that it worked well, having never tried it. But, when by a falling apple an immutable law was discovered, we gave it the crown of science, which is incontrovertible and capable of demonstration : hence that was wisdom and truth. When from the evidence of the senses my reason takes cognizance of truth, although it may appear in quite a miraculous view, I must acknowledge that as science which is truth uninvestigated. Hence the following demonstration : —

"Three weeks since I quitted my nurse and sick-room *en route* for Portland. The belief of my recovery had died out of the hearts of those who were most anxious for it. With this mental and physical depression I first visited P. P.

Quimby; and in less than one week from that time
I ascended by a stairway of one hundred and
eighty-two steps to the dome of the City Hall, and
am improving *ad infinitum*. To the most subtle
reasoning, such a proof, coupled, too, as it is
with numberless similar ones, demonstrates his
power to heal. Now for a brief analysis of this
power.

"Is it Spiritualism? Listen to the words of
wisdom. 'Believe in God, believe also in me; or
believe me for the very work's sake.' Now, then,
his works are but the result of superior wisdom,
which can demonstrate a science not understood:
hence it were a doubtful proceeding not to believe
him for the work's sake. Well, then, he denies
that his power to heal the sick is borrowed from
the spirits of this or another world; and let us
take the Scriptures for proof. 'A kingdom divided against itself cannot stand.' How, then,
can he receive the friendly aid of the disenthralled
spirit, while he rejects the faith of the solemn
mystic who crosses the threshold of the dark unknown to conjure up from the vasty deep the awe-
struck spirit of some invisible squaw?

"Again, is it by animal magnetism * that he
heals the sick? Let us examine. I have employed electro-magnetism and animal magnetism,
and for a brief interval have felt relief, from the

* Mrs. Eddy now states that Dr. Quimby was an "ignorant mesmerist" and
"magnetic healer." See Dr. Quimby's own statement, "The Philosophy of
P. P. Quimby," p. 37.

equilibrium which I fancied was restored to an exhausted system or by a diffusion of concentrated action. But in no instance did I get rid of a return of all my ailments, because I had not been helped out of the error in which opinions involved us. My operator believed in disease independent of the mind; hence, I could not be wiser than my teacher. But now I can see dimly at first, and only as trees walking, the great principle which underlies Dr. Quimby's faith and works; and just in proportion to my light perception of truth is my recovery. This truth which he opposes to the error of giving intelligence to matter and placing pain where it never placed itself, if received understandingly, changes the currents of the system to their normal action; and the mechanism of the body goes on undisturbed. That this is a science capable of demonstration becomes clear to the minds of those patients who reason upon the process of their cure. The truth which he establishes in the patient cures him (although he may be wholly unconscious thereof); and the body, which is full of light, is no longer in disease. At present I am too much in error to elucidate the truth, and can touch only the key-note for the master hand to wake the harmony. May it be in essays instead of notes! say I. After all, this is a very spiritual doctrine; but the eternal years of God are with it, and it must stand firm as the rock of ages. And to many a poor sufferer may it be found, as by me, 'the shadow of a great rock in a weary land.' * "

It will be observed, by noting the foregoing statements closely, that the lady did not understand that disease is a state of mind and the truth is its cure until this experience with Quimby took place; and it will be seen how rapidly, during the three weeks' experience referred to, she had been grasping that truth, and seeing that it was a true science, and that it was curing herself. It is now easy to see just *when* and just *where* she "discovered Christian Science."

The day following the publication of her article, it was criticised by the Portland *Advertiser;* and she then wrote a second article, replying to the criticism. In it appeared the following paragraph, referring to Quimby and his doctrine: —

" P. P. Quimby stands upon the plane of wisdom with his truth. Christ healed the sick, but not by jugglery or with drugs. As the former speaks as never man before spake, and heals as never man

* Compare this statement of Dr. Quimby's mental methods with the same writer's statements in the *Christian Science Journal,* June, 1887: " I never heard him intimate that he healed disease mentally; and many others will testify that, up to his last sickness, he treated us magnetically,— manipulating our heads, and making passes in the air while he stood in front of us. During his treatments I felt like one having hold of an electric battery and standing on an insulated stool. His healing was never considered or called anything but Mesmerism. I tried to think better of it, and to procure him public favor. He was my doctor, and it wounded me to have him despised. I believe he was doing good; and, even now, knowing as I do the harm in his practice, I would never revert to it but for this public challenge. I was ignorant of the basis of animal magnetism twenty years ago, but know now that it would disgrace and invalidate any mode of medicine."

healed since Christ, is he not identified with truth, and is not this the Christ which is in him? We know that in wisdom is life, 'and the life was the light of man.' P. P. Quimby rolls away the stone from the sepulchre of error, and health is the resurrection. But we also know that 'light shineth in darkness, and the darkness comprehended it not.' "

These excerpts are in plain language, and they speak for themselves.† The statements are made with too evident an understanding of their truth to be doubted or questioned, or afterward reversed in any particular. It should be borne in mind that your speaker was there at the time, and was familiar with all the circumstances she relates and the views expressed. The devoted regard the lady formed for her deliverer, Quimby, and for the truth he taught her, which proved her salvation, was continued to be held by her from this time (the autumn of 1862) up to a period at least four years later; for in January, 1866, Quimby's death occurred, and on February 15 she sent to me a copy of a poem she had written to his memory, and accompanied it by a letter commencing in these words: "I enclose some lines of mine, in memory of our much-loved friend, which, per-

† Mrs. Eddy's letters to Dr. Quimby are even more emphatic in their praise of him, and leave no doubt in the reader's mind that she believed his methods to be of the highest spiritual character. I have quoted from these letters in *The Arena*, May, 1899.

haps, you will not think overwrought in meaning:
others must, of course."*

The poem, which had been printed in a Lynn
newspaper, is as follows:—

LINES ON THE DEATH OF DR. P. P. QUIMBY, WHO HEALED
WITH THE TRUTH THAT CHRIST TAUGHT, IN CONTRA-
DISTINCTION TO ALL ISMS.

Did sackcloth clothe the sun, and day grow night,
  All matter mourn the hour with dewy eyes,
When Truth, receding from our mortal sight,
  Had paid to error her last sacrifice?

Can we forget the power that gave us life?
  Shall we forget the wisdom of its way?
Then ask me not, amid this mortal strife,—
  This keenest pang of animated, clay,—

---

* This letter is reprinted below at greater length:—

LYNN, Feb. 15, 1866.

MR. DRESSER:

Sir,— I enclose some lines of mine in memory of our much-loved friend,
which perhaps *you* will not think overwrought in meaning: *others* must, of course.

I am constantly wishing that *you* would step forward into the place he has
vacated. I believe you would do a vast amount of good, and are more capable of
occupying his place than any other I know of.

Two weeks ago I fell on the sidewalk, and struck my back on the ice, and was
taken up for dead, came to consciousness amid a storm of vapors from cologne,
chloroform, ether, camphor, etc., but to find myself the helpless cripple I was
before I saw Dr. Quimby.

The physician attending said I had taken the last step I ever should, but in
two days I got out of my bed *alone* and *will* walk; but yet I confess I am
frightened, and out of that nervous heat my friends are forming, spite of me, the
terrible spinal affection from which I have suffered so long and hopelessly. . . .
Now can't *you* help me? I believe you can. I write this with this feeling: I
think that I could help another in my condition if they had not placed their intelligence in matter. This I have not done, and yet I am slowly failing. Won't you
write me if you will undertake for me if I can get to you? . . .

Respectfully,   MARY M. PATTERSON.

To mourn him less: to mourn him more were just,
 If to his memory 'twere a tribute given
For every solemn, sacred, earnest trust
 Delivered to us ere he rose to heaven.

Heaven but the happiness of that calm soul,
 Growing in stature to the throne of God:
Rest should reward him who hath made us whole,
 Seeking, though tremblers, where his footsteps trod.

MARY M. PATTERSON.

LYNN, Feb. 22, 1866.

Oh that in an evil hour she had never been tempted to erase the sentiments of that poem, which was *not* an overwrought tribute to the memory of our much-loved friend! *

But let that charity which rejoiceth not in iniquity, but rejoiceth in the truth, and never faileth, not fail even here; for we have the truth to rejoice in. This truth which P. P. Quimby brought forth, and for years labored so unceasingly to give to the world, and finally laid down his life in its cause, — this glorious truth is still blessing us; and it will do so more and more unto the perfect day. It is a revelation of truth that makes us free indeed! And we have only to set aside self-love and self-glory and work earnestly in this cause, by every word and deed of love that opportunity offers, to find ourselves growing gradually into all wisdom and understanding, and out

---

* The first edition of "Science and Health" also contained a tribute to Dr. Quimby. But this passage was afterward omitted and the edition suppressed.

of and away from every ill and every form of
unhappiness in existence.*

I cannot better close this address than by quot-
ing an extract from one of P. P. Quimby's manuscripts,† which will show the spirit of the man,
as well as give an indication of the truth he
preached : —

"Every disease is the invention of man, and
has no identity in wisdom; but, to those who
believe it, it is a truth. If everything man does
not understand were blotted out, what is there
left of man ? Would he be better or worse, if
nine-tenths of all he thinks he knows were blotted
out of his mind, and he existed with what is true?

" I contend that he would, as it were, sit on the
clouds, and see the world beneath him tormented
with ideas that form living errors, whose weight is
ignorance. Safe from their power, he would not
return to the world's belief for any consideration.

---

*I have quoted Mrs. Eddy's own words to the effect that (1) she began to
discover the science of mind healing in 1844; (2) the excerpts from the *Courier*
and her letters to Dr. Quimby prove that she *acquired this science when with
Dr. Quimby* ; (3) now let us read her own comments (*Christian Science Journal*, June, 1887): "Did I write those articles purporting to be mine? I might
have written them, twenty or thirty years ago, for I was under the mesmeric
treatment of Dr. Quimby from 1862 until his death in 1865. He was illiterate
and I knew nothing then of the Science of Mind-healing, and I was as ignorant
of mesmerism as Eve before she was taught by the serpent. Mind science was
unknown to me; and my head was so turned by animal magnetism and will-
power, under his treatment, that I might have written something as hopelessly
incorrect as the articles now published in the Dresser pamphlet." See Appen-
dix A.

† Dr. Quimby's manuscripts were revised and copied under his direction by
the Misses Ware and by Mr. George A. Quimby. They were written in
1859-65.

"In a slight degree, this is my case. I sit as it were in another world or condition, as far above the belief in disease as the heavens are above the earth, and, though safe myself, I grieve for the sins of my fellow-man; and I am reminded of the words of Jesus when he beheld the misery of his country-men : 'O Jerusalem, how oft would I have gathered thee, as a hen gathereth her chickens, but ye would not!'

"I hear this truth now pleading with man, to listen to the voice of reason. I know from my own experience with the sick that their troubles are the effect of their own belief,— not that their belief is the truth, but their beliefs act upon their minds, bringing them into subjection to their belief, and their troubles are a change that follows. ·

"Disease is a reality to all mankind; but I do not include myself, because I stand outside of it, where I can see things real to the world and things that are real to wisdom. I know that I can distinguish that which is false from a truth, in religion or in disease. To me, disease is always false; but, to those who believe it, it is a truth, and the errors of religion the same. Until the world is shaken by investigation, so that the rocks and mountains of religious error are re-moved and the medical Babylon destroyed, sick-ness and sorrow will prevail. Feeling as I do, and seeing so many young people go on the broad road to destruction, I can say from the bottom of

my soul: O Priestcraft! fill up the measure of
your cups of iniquity; for on your head will come,
sooner or later, the sneers and taunts of the people. Your theory will be overthrown by the voice
of wisdom, that will rouse the men of science, who
will battle your error, and drive you utterly from
the face of the earth. Then there will arise a
new science, followed by a new mode of reasoning, which shall teach man that to be wise and
well is to unlearn his errors." *

* This article, like many others, suggests a depth of understanding beyond
what Dr. Quimby could explain to his followers. It was only to a very few that
he tried to communicate his highest insights. Consequently, when outsiders
undertook to state his belief, and when Mrs. Eddy's followers nowadays have
undertaken to tell what he believed, there was naturally a wide divergence of
opinion. In his own day he was frequently called a Spiritualist, quack, mesmerist, etc.; but, to know what he truly was, one must know him intimately, as did
my father and those who copied his manuscripts, and in some degree Mrs. Eddy
herself, as shown by her letters. One of these intimate followers informs me
that "very often after a sitting with a patient the patient would say, 'O doctor,
tell me how you cure?' Usually he would say, 'Oh, I don't know myself,'
simply to get rid of the patient. How could he in one answer, in a few minutes'
time, tell one, who had not the remotest conception of his method, what he had
been years studying? . . . But there is plenty of written evidence from himself
just what he did believe, and what others *say* he believed, is no evidence at all,
but simply their opinion."

APPENDIX A.

In order that the reader may judge for himself, the following letter to the editor of the Boston *Post* is reprinted as a further illustration of Mrs. Eddy's statements : —

We give our first leisure to reply to the false allegements appearing in a letter of J. A. Dresser in your issue of the 24th ult., and are able to prove the statements relative to those allegements hereinafter made. While founding what is new and abstract, such as Christian Science, truths revolutionary in character blessing mankind, but not understood at the period in which they appear, ignorance and malice have thrown in our way implements of their own calculated to retard our work. Notwithstanding all this, since 1866 we have advanced steadily in introducing into Massachusetts the science of mental healing.

We had laid the foundations of mental healing before we ever saw Dr. Quimby ; were an homœopathist without a diploma, owing to our aversion to the dissecting-room. We made our first experiments in mental healing about 1853,* when we

---

* In the *Christian Science Journal*, June, 1887, Mrs. Eddy gives the date as 1844. In "Retrospection and Introspection" (page 28) she says, "It was in Massachusetts, in the year 1866, that I discovered the Science of Divine Metaphysical Healing, which I afterward named Christian Science." Again (page 51), "In 1867 I introduced the first purely metaphysical system of healing since apostolic days." This she named "the great discovery," on a "basis so hopelessly original" that she charges others with plagiarisms from "the precious book," — "Science and Health," — "the only known work containing a *correct* and *complete* statement of the Science of Metaphysical Healing, its principles and practice."

were convinced that mind had a science which, if understood, would heal all diseases. We were then investigating that science, but never saw Dr. Quimby until 1862. Mr. Dresser's statement that "Mrs. Eddy knows positively that the assertions of E. G. in last Monday's *Post* are a tissue of falsehoods" is untrue. We answer for all times that those assertions were strictly true. We never were a student of Dr. Quimby's, and Mr. Dresser knows that. Dr. Quimby never had students, to our knowledge. He was a humanitarian, but a very unlearned man. He never published a work in his life; was not a lecturer or teacher. He was somewhat of a remarkable healer, and at the time we knew him he was known as a mesmerist. We were one of his patients. He manipulated his patients, but possibly back of his practice he had a theory in advance of his method; and, as we now understand it, and have since discovered, he mingled that theory with mesmerism. We knew him about twenty years ago, and aimed to help him. We saw he was looking in our direction, and asked him to write his thoughts out. He did so, and then we would take that copy to correct, and sometimes so transform it that he would say it was our composition, which it virtually was; but we always gave him back the copy, and sometimes wrote his name on the back of it. We defended Dr. Quimby from unmerited scorn, asserted in public that his practice was not mesmerism; for we so believed it then, being utterly ignorant of the nature, theory, or practice of mesmerism. Since then the sin and subtlety of a student who departed from our teachings, and became a malpractitioner, caused us to investigate the subject of mesmerism, when we learned that manipulation includes animal magnetism; and, if one manipulates the sick, no matter what his theory is, it precludes the possibility of his practice being mental science. And, if he understands mental healing and its science, he will see that manipulation retards healing instead of helps it. We have no doubt that Dr. Quimby's motives were good, for we understood him to be a moral man.*

The malpractitioners, whose hidden crimes we have endeavored to expose, may put the burden of introducing plagiarisms, to stop the circulation of our books, on the shoulders of the new party, — namely, J. A. Dresser; but they cannot hide the malice aforethought through which they are seeking to wrest from us the public confidence, and so disarm our ability to warn and forearm the people against what we have seen of their crimes,— the danger of mesmerism and its power to kill some individuals. Nor can they silence many witnesses to some of their mental murders, which the general ignorance of this subject has hitherto prevented being duly investigated. It is not many years since one of these malpractitioners prosecuted another one for attempting, through mesmerism, to destroy the life of a lady in Ipswich. But the plaintiff in that case, since accused of the same crime, now avoids the question of mesmerism and malpractice. Step into Mr. Aren's office on Chester Park, and you can obtain some advertising pamphlets gratuitously, which, by comparison, are found to contain verbatim paragraphs from our work, published in 1870; and he studied for the first time mental healing of my husband in 1879.

* It may be asked, What was Dr. Quimby's opinion of Mrs. Eddy? He paid no heed at first to the prophecy of a friend that she would "steal his ideas and set up for herself," but afterward declared that she had "no identity in honesty."

The private letter from a lady which the gentleman (?), Mr. Dresser, has on exhibition, was written under the following circumstances : —

At Swampscott, Mass., in 1866, we recovered in
a moment of time from a severe accident, considered fatal by the regular physicians, and re-
gained the internal action that had stopped and
the use of our limbs that were palsied. To us
this demonstration was the opening of the new
era of Christian science. We then gained a proof
that the principle, or life of man is a divine intelligence and power, which, understood, can heal
all diseases, and reveals the basis of man's immor-
tality. But the minds around us at that time
were unacquainted with our mental theory. One
individual of strong intellectual power and little
spirituality even occasioned us some momentary
fears of our ability to hold on to this wonderful
discovery. In one of these moments of fear we
wrote Mr. Dresser, but we wrote him after we had
proven our ability to work out the problem of
mental healing. The failing state referred to was
a state of mind ; and there are living witnesses to
our health at that time,— we were never as well
before in our life. It was but a timid hope that we
referred to,— a trembling explorer in the great
realm of mental causation, where evil is more
apparent and good more divine. We sought for
once the encouragement of one we believed
friendly, also with whom we had conversed on
Dr. Quimby's method of healing ; and, when we
had said to him, "It is a mystery," he replied to
the effect that he believed no one but the doctor
himself knew how he healed. But, lo ! after we
have founded mental healing and nearly twenty
years have elapsed, during which we have taught

some six hundred students and published five or six thousand volumes on this subject, already circulated in the United States and Europe, the aforesaid gentleman announces to the public Dr. Quimby, the founder of mental healing.

In 1862 my name was Patterson; my husband, Dr. Patterson, a distinguished dentist. After our marriage I was confined to my bed with a severe illness, and seldom left bed or room for seven years, when I was taken to Dr. Quimby, and partially restored. I returned home, hoping once more to make that home happy, but only returned to a new agony,— to find my husband had eloped with a married woman from one of the wealthy families of that city, leaving no trace save his last letter to us, wherein he wrote, "I hope some time to be worthy of so good a wife." I have a bill of divorce from him, obtained in the county of Essex. My first husband * was Colonel Glover, of Charleston, S.C. Six months after our marriage he died of scarlet fever. Our only child was born six months after his death. To our brief happy union and to the noble character of my husband there are tender testimonials and resolutions, passed by the brother Masons of St. Andrew's Lodge, in which he had taken the degree of "Royal Arch Mason," which articles were published in the *Freemasons' Magazine*, edited at that time by Charles Moore, of Massachusetts.

We shall not descend to notice any further falsehoods through the press, since there is so much good we can do we cannot afford to sacrifice our time.

No. 569 COLUMBUS AVENUE,
    March 7, 1883.

* An earlier husband bore the name of Morse, not "Mason" as erroneously stated in *The Arena*, May 1899.

THE following statement by Mr. A. J. Swartz, who made a careful investigation of the facts in 1888, is of particular value, since it comes from an outsider who had no particular reason to defend Dr. Quimby.

In an editorial in the *Mental Science Magazine* of April, 1888, Mr. Swartz says : —

If, through the mistakes of ambition, the writer of " Science and Health " has led many of us to ignore for a time the good works of other true reformers, our timely rescue from unchristian and uncharitable ways is secured by the palpable mistakes she and her immediate following have made and are still making.

Her claim that she was the founder of mind-healing is false; nor is it true that she first conceived the idea of applying Christian to the science. Having been misled in the past, we have done wrong toward *facts* by claiming to our readers that she first designated it as above; and hence it is our duty to correct the error.

All who know Julius A. Dresser, of Boston, know him to be a truthful and most excellent gentleman. Some of the Portland friends remember Mr. Dresser when he was with Dr. Quimby in their city in 1860. He tells me personally that he has seen Dr. Quimby's early writings, in which are seen by his own words the origin of the idea or term " Christian Science." This statement was fully corroborated to me by a lady now of Boston, the daughter of Judge Ware, prominent as United States judge. This lady and her sister resided at Portland, and were both among the earliest patients the doctor treated. They assisted him in

the records and correspondence as his scribes.
This lady says that the doctor was often invent-
ing, that he was a jeweller. She had seen a
clock that he invented. He had also been a mesmerist. She says that he wrote and kept manuscripts which he named "Science of Health and
Happiness," that she also often wrote sentiments
and sayings he uttered among them. She says
that she heard it remarked repeatedly by all of
them, when talking of the wonderful cures by the
doctor, that "this is a new life and the real science
of Christianity." Here, then, we have in the con-
versation of the doctor and of those working with
him the origin of the term Christian Science.
Those who first called it the "Science of Christianity," in the United States Hotel, before the
author of "Science and Health" visited Dr.
Quimby at the International Hotel for cure in
1862, were the originators of this title.

The lady referred to above showed me a printed
circular which bears the name of Dr. P. P. Quimby
when he was in the active healing work. His
method of diagnosis is set forth in this circular,
and it is purely spiritual; for he sat with eyes
closed, and arrived at their "beliefs" of disease, as
he called them, by impression. He says he could
cure those whose troubles or diseases he described
correctly, while he was not successful as a rule
with those whose fears or beliefs he could not
discern ; nor did he charge such.

It is evident from his own circular that his
process of diagnosis was that which is to-day
called intuition, thought-transference, or spiritual
perception. His method of curing was to change
the patient's beliefs and fears, which he could
usually do by assuring them orally and mentally

of success. Here, then, we have the true mind healing, as Dr. Evans and other correct writers present it. To contradict orally increases confusion ; and to contradict mentally, as some teach, does not cure disease.

In the same issue of the *Mental Science Magazine* Mr. Swartz says : —

To check the spread of recent facts from Portland, the author of "Science and Health" sent from Boston over her signature to the Portland *Daily Press*, while I was there, a pay article called "IMPORTANT OFFER." In it are four provisos, hence too metaphysical to be a fair offer or in the least liable to endanger any outlay. While it offers to pay the costs of printing the Dr. Quimby manuscripts, it agrees, among other provisos, to do so, "*provided* that I am allowed first to examine said manuscripts, and that I find they were Mr. P. P. Quimby's own compositions, and not mine that were left with him many years ago, or that they have not since his death, in 1865,"— he died in January, 1866,— "been stolen from my published works," etc.

The "offer" hinges fully upon her own decision, and demands her possession first of the manuscripts. Knowing that the owner would not permit this, it was safe to make a provisional offer. It will always be.

I then wrote to George A. Quimby, who is an honest clothing manufacturer in Belfast, Me., and sent the clipping "Important Offer," from which we extract above. I desired to know whose claims are truthful, for I intend ever to defend such only. I asked various questions. On

Feb. 22, 1888, he replied in a very gentlemanly manner; and from this reply we extract:—

"Your letter with enclosure at hand. I judge that you offer to defend the memory of my father, the late P. P. Quimby. . . . Please permit me to say that I have no doubt of your kind intention to come to the rescue of my father, but I do not feel that there is the slightest necessity for it. . . If I were in prison, in solitary confinement for life, I should be too busy to get into any kind of a discussion with Mrs. Eddy.

"I have my father's manuscripts in my possession, but will not allow them to be copied nor to go out of my hands. Answering your further inquiries, I have no written article of Mrs. Eddy's in my possession, have never had, nor did my father ever have any, nor did she ever leave any with either of us. Neither of us have ever 'stolen' any of her writings nor anything else. In fact, we both have been able to make a living without stealing.

"I expect to have an article in the March number of the *New England Magazine* * on my father and his work. I think that perhaps you will see in that what you desire.

"Yours truly, Geo. A. Quimby."

I have read the article in the *New England Magazine,* published at Boston, Mass. It is quite extended, and will interest many. I can make no quotation from it, as I have no access now to it. I quote, however, a few lines extracted by another "from the manuscript of Dr. P. P. Quimby, Portland, Me." These extracts are in the posses-

---

* The substance of this article is reprinted, "The Philosophy of P. P. Quimby," p. 11.

sion of Mary Lyman Storrs, Albany, N.Y., and read as follows : —

"Disease being made by our belief, or by our parents' belief, or by public opinion, there is no formula to be adopted, but every one must be reached in his particular case. Therefore, it requires great shrewdness or wisdom to get the better of the error. Disease is our error and the work of the devil, who is the father of all falsehoods, of whom Christ, or Truth, saith, 'When he speaketh of a lie, he speaketh of his own, for the truth abode not in him.' But, happily, he hath his cloven foot; and, if you are as wise as your enemies, you will get the case.

"I know of no better counsel than Jesus gave to his disciples when he sent them forth to cast out devils and heal the sick, and thus in practice preach the truth; viz., 'Be ye wise as serpents and harmless as doves,'— i.e., never get into a passion, but in patience possess ye your souls. At length ye weary out the disease, and produce harmony by your truth destroying their error. Then it is you get the case. Now, if you are not afraid to face the error and argue it down, then you can heal the sick. When wisdom, which is Truth, calls upon you, 'Adam, where art thou?' and you are afraid and hide yourself away, then you cannot heal the sick with Truth."

While in Portland recently, I visited the large room in the United States Hotel used by Dr. Quimby as his healing-room. He also occupied rooms at the International Hotel, for he cured for several years in Portland.

In this hotel one Mr. Rogers, a prominent business man of Portland, told me that he was born in this city in 1827; that he knew Dr.

Quimby well. He gave me quite a history of the "strange little man" and his methods of treatment. Dr. Quimby treated this gentleman for severe troubles some three and a half years in this hotel. He would tell the doctor that he was "a fraud or a humbug for attempting to cure disease without drugs." The doctor, who was a peculiar jester, would turn it with some joke that, if he was a fraud, "the people and you will come to be cured." He says it must be admitted that he had many patients and performed wonderful cures. My inquiries relating to the power, agency, or cause of his cures, only corroborated the same that we have often heard; viz., that the doctor did not profess to know what it was other than Truth or Mind-power. Whatever he could do to act on the patient's mind,— *i.e.*, his beliefs or fears,— so as to restore hope or confidence, would result in curing. When asked why he put his hands in water and then on the patient's head, he said he did so with part of them to stimulate their hopes, and so that they would feel that something was being done for them. The doctor often said that a science would come out of it, and he hoped others would bring it out better. His instrumentality was honored, and his predictions have been fulfilled. If he used clear water on the brow of some God sent to him, he also cured many with Mind, with Truth, with mental treatment.

THE *Christian Science Sentinel* of Feb. 16, 1899, prints the following letter to Mrs. Eddy : —

It might be interesting to you to know that Mr. A. J. Swartz, of Chicago, went to see the late Dr. P. P. Quimby's son, and procured his father's writings for the purpose of having them published in order to show the world that your ideas were borrowed from Quimby. After having examined them, to their utter disappointment, it was found there was nothing that could compare in any way to " Science and Health "; and he, Swartz, concluded that it would aid you too much to publish them, so they were returned to the owner. Mrs. Swartz saw and read these manuscripts, and she gave me this information.

MARY H. PHILBRICK.

AUSTIN, ILL., May 18, 1892.

The truth is that neither Mr. Swartz nor Mrs. Swartz ever saw Dr. Quimby's manuscripts. Mr. Quimby read a few paragraphs from his father's manuscripts to Mr. Swartz, but did not permit him to examine them. *Mrs. Swartz was not present at all.*

In the *Mental Science Magazine* of June, 1888, Mr. Swartz says : —

During the latter part of the month I went on to Belfast, near 150 miles to the north and east from Portland. Much has been said relative to the "Quimby Manuscripts," so called. We have heard that Dr. Quimby left various writings setting forth his views and practice in mental phenomena and the healing of disease by a method which he called his "system," or "theory." It has been claimed that he cured multitudes of individuals, and that from his discoveries and practice the great system of mental or modern Christian healing came forth. It has been said repeatedly that Dr. Quimby intended to publish a work showing the origin of mind-healing in this nation, but that his decease in January, 1866, prevented its appearance. It has been declared often that his son, George A. Quimby, a clothing manufacturer for many years in Belfast, intends to publish to the world a book of his father's discoveries and practice that will show the successful healing which he began prior to 1850, and in which he continued until 1865.

It is understood that one who was Dr. Quimby's patient in the summer of 1862 claims to have discovered mental healing, and to have originated all the science. It is also strongly affirmed that the aforesaid writings will show that all the foundation principles of mental healing, as well as the terms used to express them, also that disease, as defined by all science healers, will be found in the written views of the said Dr. Quimby years before any other wrote upon the subject.

In these issues, facts and good faith only will govern sincere workers in this field. That others took up his mantle and made many of his discoveries public, which have so blessed the race, can be regarded as a noble work only; but the question of fairness and good faith is, Have any others used and cultivated his original ideas, words, definitions, and discoveries under the claim that they were the first discoverers of this system ?

Feeling that honest hearts are wounded by those who deny that human beings have hearts, and knowing that too much of high principle is at stake to disregard the truthfulness of the varying claims, I felt led of the Spirit to go and ascertain facts about the origin of this mental system, the said manuscripts, etc.

I found George A. Quimby to be a man far superior to what I had supposed. He is the busy proprietor of an industry that he has pursued some sixteen years. He, together with several others, assisted Dr. Quimby for years in the details of his healing practice, such as the receiving of patients, recording names, and aiding him as the cause required. During these years of practice the doctor was written up by many editors. He also wrote at various times his ideas and discoveries. Frequently he would utter many valuable truths in his instructions or conversations with others. His views were often written by those associated with him, and then submitted to him for approval or correction. These writings by himself, and by those in his employ for years who wrote for him, constitute the said manuscripts. All literary people know that manuscripts proper consist of the views written by an individual, or his views penned by another for him, provided that he corrects, approves, and appropriates the same. Through both of these proper methods extended manuscripts of Dr. Quimby's system and discoveries were kept by himself and his family; for he intended to publish a work from them, and doubtless would have done so, had he lived a little longer. His son, and several other truthful persons, agree that the facts relating to his manuscripts are as I give them herein.

I also called upon an elder son, cashier of the chief bank in Belfast. I found that Dr. Quimby was highly respected in his day and work. This I learned from various reputable citizens. He had a name for modesty and strict honesty that was better than riches. He never professed scholarly erudition, as the world counts it; but he was a student and an investigator, who reached the inner workings of the human mind. He was the wiser in this because of having been a successful mesmerist fifty years ago. Mesmerism is mental phenomena, or mind acting in or upon mind. Mesmerism in itself is not an evil. Every mental scientist should know that motive or intention gives coloring or character to every deed.

I saw a perfect clock in George's store, invented and made by his father many years ago; for he was the instrument through whom discoveries aside from mental Christian healing came to humanity.

Now the question is, Does George A. Quimby possess the said manuscripts, and does he intend to publish them? He most certainly does possess them; and I hope that he may decide to bring out a book from them, or at least to publish his father's views relating to disease, in what it consists, how it was handled, how he diagnosed disease, and his personal experiences which led to the discovery of disease as a mental condition,— as beliefs, fear, wrong thinking, etc.,— and then how he removed these states by instructing or by explaining Truth to the patient. All these ideas were held and practised by Dr. Quimby for more than twelve years before any writer or author in the system of mental healing published any work on the subject. "Mental Cure," by Dr. Evans, was pub-

lished in 1869. " Science and Health " was published in 1875. The writings, as written and
indorsed by Dr. Quimby, were by him called
" Science of Health and Happiness." These are
transcribed into blank books, which I saw. Another interested person has a certified copy of the
transcribed writings to insure against accident by
fire. Hence it is evident that a deep interest is
manifested in them. Also the original writings
are there.

*Mr. Quimby read to me enough from the said
manuscripts to fully justify all I state above, relating to his father and to disease; also to convince
me that to him were revealed all the essential claims
or fundamental principles relating to disease direct,
as now held by every author in mental or Christian
science.* This can be established, and I venture
the statement that it will be. I saw and heard
letters read that were written to Dr. Quimby
early in 1862, which show that facts are far from
the claims set forth elsewhere. I have been misled by dishonesty and ambition. Many letters
and these writings are preserved by Mr. Quimby
in his safe. He loaned me for several days a
book of 116 pages, consisting of printed articles
by editors, by patients, and the immediate friends
of Dr. Quimby; also, some written by the doctor.
These were taken from newspapers, and bear
printed dates reaching back to 1840. Dr.
Quimby's signature is printed as the author of
some of these. They are all interesting and
valuable. He kindly permitted me to copy all of
these (newspaper) articles I desired for publication, or for use in my own forthcoming work of
interpretation, or New Era doctrines. These will
be of deep interest, and every line will be credited

to the respective writers and papers of those years. If I use them, they will be a department by themselves.

Although some of the newspaper articles I copied were by Dr. Quimby, still the manuscripts are the property of George A. Quimby, who may publish them at some future time, when he feels an interest sufficient to call them out. It may be difficult for scientists to see that he is not interested in the cause, that he is in no way engaged in it, but is attending only to business interests, and pays very little attention to the claims or desires of others relating to the writings. *Various persons have tried to possess them, but without avail.*

Whenever Mr. Quimby decides to publish them, he has both the brains and the means. He is a member of the Maine Press Association, and is a writer of fine ability. He is a kind-hearted gentleman, and has given considerable time to the employ of several first-class papers,— one, the Boston *Globe*, as travelling correspondent.

For the *Courier*.

SONNET.

Suggested by Reading the Remarkable Cure of Captain J. W. Deering *

TO DR. P. P. QUIMBY.

'Mid light of science sits the sage profound,
Awing with classics and his starry lore,
Climbing to Venus, chasing Saturn round,
Turning his mystic pages o'er and o'er,
Till, from empyrean space, his wearied sight
Turns to the oasis on which to gaze,
More bright than glitters on the brow of night
The self-taught man walking in wisdom's ways.
Then paused the captive gaze with peace entwined,
And sight was satisfied with thee to dwell;
But not in classics could the book-worm find
That law of excellence whence came the spell
Potent o'er all,— the captive to unbind,
To heal the sick and faint, the halt and blind.

MARY M. PATTERSON.

* Printed from the original manuscript by permission of George A. Quimby.

APPENDIX B.

WHEN "The True History of Mental Science"
first appeared, there was, of course, strenuous
effort to contradict its statements. The gentleman who formerly revised Mrs. Eddy's manuscripts has since told me that Mrs. Eddy said the
pamphlet must be answered. Accordingly, the
gentleman mentioned point after point, asking if
father's statements were correct. Mrs. Eddy
could not deny them. Then there is nothing to
say, was the rejoinder. However, the *Christian
Science Journal* of June, 1887, contained an emphatic denial, to which father, when urged, made
the following reply : * —

A CARD.

*To All Christian Scientists, Mental Scientists, and Meta-
physicians everywhere.*

The lecture recently issued in pamphlet form,
entitled "The True History of Mental Science,"
was written and delivered in response to a call for
facts relating to the discovery of the principle and
the founding of this science. The lecture itself

* *Mental Healing Monthly*, July, 1887.

shows that I could have given the facts years sooner; but, having no desire to oppose or antagonize any person whatever, I withheld the facts until the demand for them became great enough to show that the publishing was for the cause of the truth only. Since the publication, Mrs. M. B. G. Eddy has attacked me, in the June number of her *Christian Science Journal*,* for telling these facts. But, as readers of the two statements, hers and mine, do not themselves know which tells the truth, I shall ask them to compare the two, and judge for themselves, not only of the statements made, but of the spirit manifested by the two writers. Read, and decide for yourselves which one is humbly trying to serve the cause of truth, and which has some other motive. I have no controversy with anybody, and shall not reply to Mrs. Eddy. I have no personal cause to maintain, either for myself or Dr. Quimby. All is for the simple truth and the real facts. And I have no fear that the truth will suffer, or fail to be finally vindicated.   JULIUS A. DRESSER.

---

* Another attempted denial appeared in the Boston *Traveler*, April 25, 1899, in which it was claimed that the entire controversy was settled in court in 1883. But the suit then brought related to a personal quarrel between Mrs. Eddy and Dr. E. J. Arens, who claimed that "Science and Health" was the work of Dr. Quimby, and had no bearing on the real point at issue. No one who knows the facts now claims that Dr. Quimby wrote Mrs. Eddy's book. Dr. Arens obtained his information at second hand. The court ruled that Dr. Arens could not legally publish a pamphlet in which he (Dr. Arens) had restated Mrs. Eddy's ideas. But *nothing was or could be settled in regard to Mrs. Eddy's indebtedness to Dr. Quimby, because neither the court nor Dr. Arens possessed the Quimby manuscripts.* The claims of the *Traveler* are therefore without foundation.

APPENDIX C.

SUCH wide-spread fear has been elicited in the "malicious animal magnetism" of the Eddy school that I venture to reprint an article of mine written for *The Life*, Kansas City, Jan. 22, 1896 : —

MENTAL PROTECTION.

The study of mind in its relation to health adds a new terror to an already plentiful supply of fears, unless the study be sufficiently profound to reveal the one great law which the new philosophy of health everywhere makes prominent. It becomes evident to the most superficial observer that mental influence may be maliciously used ; and the new philosophy, in fact, throws a flood of light on the subtle sources of crime to which the masses of mankind are in general subjected. In Belgium and other foreign countries it has been found necessary to limit the use of hypnotism by law. In this country the use of hypnotism in crime has thus far received little attention, although its possible effects have already become the subject of serious inquiry.

It is well known by those who have investigated the new philosophy that it aims to be in strict accordance with Christianity at its best. There is thus nothing to fear from any genuine mental healer, for the followers of the new thought are making a most exemplary effort to realize in actual life the law of love which Jesus taught both by precept and by example.

But the mental healing world is split into two camps,— those who consistently carry out the law of love and those whose motive it is to support their leader at any cost. Concerted treatment to ruin the prospects and the business of those who fearlessly and charitably publish the truth about that leader is now a widely known and most deplorable fact. People hesitate to tell what they know lest they become the object of this underhand and most pernicious practice. They fear for themselves and their friends. And thus aspersions are cast on the whole society of earnest truth-seekers whose sole object it is to win and promulgate impersonal truth.

What is the law which governs the entire range of mental influences and telepathic communications? Is it not the law that like attracts like? Is contamination possible where there is no receptivity?

"According to thy faith be it done unto thee," is the law enunciated throughout the New Testament. We partake of the Spirit, and grow strong in its ministry in proportion to our receptivity, our openness and obedience to it. In return for obedience it affords a protection which no words can measure. It is a shield so secure that one may rest in love and trust without giving a thought to one's possible enemies.

What temptation is it to a righteous man to go into the company and observe the enticements offered by the intemperate and the depraved? What point of contact is there between the unselfish philanthropist and the dishonest dealer who tries to ruin the philanthropist's schemes because his own business is at stake?

Observe society at large, and especially those aspects of it where people are drawn together in ties of friendship and co-operation,— either for good or ill,— and do you not find some natural affinity constituting their bond of union and sympathy? Even thieves may plot and conspire together according to the same law which binds man to man on the highest plane of being. But the plans, the energy, the appeals, and even the threats of one man pass by another without making the slightest impression, when there is no common ground of fellowship between them.

In the thought-world the law is yet more strikingly exemplified. Your morbid thought and fear dies out immediately, unless it finds its like, unless it meets encouragement. If you feed the ever-ready consciousness with the sensational matter of our Sunday papers, it will become more and more open to all that is deplorable in our American civilization. If you never permit your thought to dwell upon accidents, murders, medical advertisements, reports of trials, etc., you thereby grow strong in the higher direction of thought. For the mind is ever aggressive. It is always engaged upon something; and he is wise, he is healthy, he is pure and helpful to his fellow-men, whose thought is habitually centred on the good, the true, the beautiful, with that *power which recognizes no opposition*, and that trust which not all the evil-minded people and all the baneful influences in the world can prevail against.

There is nothing to fear, then, from that with which we have nothing in common. By just so much of spiritual power, just so much of true wisdom, of knowledge of ourselves as related to other minds, and just so much repose of character as we possess, are we protected, guided, and strengthened from within. Only that part of us which stoops to what is below it is ever contaminated by the temptations and corruptions of society. Where peace and trust dwell, the devil himself fears to enter.

NOTE.

The attention of the reader is called to the following *exposés* of Christian Science : —

The Arena, May, 1899, "Christian Science and its Prophetess ": (1) The Facts in the Case, by H. W. Dresser; (2) The Woman and the Book, by Josephine Curtis Woodbury.

"An English View of Christian Science: An Exposure," by Anne Harwood. Fleming H. Revell Company, New York.

"Christian Science Examined," by Henry Varley. 35 cents. Revell.

"What is Christian Science?" by Rev. P. C. Wolcott, B.D. 35 cents. Revell.

"The Comedy of Christian Science," by W. H. Mallock, *National Review*, March, 1899.

"Christian Science and its Legal Aspects," by W. A. Purrington, *North American Review*, March, 1899.

"The Passing of Christian Science," by G. C. Harger, Jr. Paper, 30 cents. D. J. Stoddard, 365 Washington Street, Buffalo, N.Y.

www.ingramcontent.com/pod-product-compliance
Lightning Source LLC
Chambersburg PA
CBHW021539270326
41930CB00008B/1305